YOUR KNOWLEDGE HAS VALUE

- We will publish your bachelor's and master's thesis, essays and papers

- Your own eBook and book - sold worldwide in all relevant shops

- Earn money with each sale

Upload your text at www.GRIN.com
and publish for free

Bibliographic information published by the German National Library:

The German National Library lists this publication in the National Bibliography; detailed bibliographic data are available on the Internet at http://dnb.dnb.de .

Imprint:

Copyright © 2013 GRIN Verlag
Print and binding: Books on Demand GmbH, Norderstedt Germany
ISBN: 9783668748347

This book at GRIN:

https://www.grin.com/document/430763

Juan Casiano

A Hypothetical Method of Attempting to Break the Current Sailing Record Around the World using Spherical Trigonometry

GRIN Verlag

GRIN - Your knowledge has value

Since its foundation in 1998, GRIN has specialized in publishing academic texts by students, college teachers and other academics as e-book and printed book. The website www.grin.com is an ideal platform for presenting term papers, final papers, scientific essays, dissertations and specialist books.

Visit us on the internet:

http://www.grin.com/

http://www.facebook.com/grincom

http://www.twitter.com/grin_com

Propose a Hypothetical Method of Attempting to Break the Current Sailing Record

Around the World using Spherical Trigonometry.

Juan Antonio Casiano
Extended Essay on Mathematics

I.D.E.A College Preparatory an I.B World School
October 30, 2013

Word Count: 2229

Abstract:

Ever since the creation of math, mathematicians have attempted to extend, or challenge the work of another mathematician with the intent to try and disprove their discoveries. The applications of math we now use to solve the problems of life, are due to discoveries of these great minds. Mathematics is no longer a system to count objects, as this examination will attempt to : **Propose a Hypothetical Method of Attempting to Break the Current Sailing Record Around the World using Spherical Trigonometry.**

The scope in which this examination will take into account is that of spherical trigonometry at its sole. Situations will be adjusted to make spherical trigonometry the tool to attempt to challenge the current record of sailing around the world. It will not include the vector components entirely. Needless to say, the majority of the trigonometry used in this examination will be explained just enough to be understandable for the common math enthusiast. The record breaking component, is only a form in which this sub-branch of spherical geometry can be applied in the real world.

The result of this examination ended with a success. The method taken resulted in breaking the current record held by Loïck Peyron within an astonishing 45 days 13 hours 42 minutes and 53 seconds. But from this examination it was derived that, with respect to the given points, that you could go around the world in 20 days 17 hours 5 minutes and 17 seconds when going at a speed of 40 knots. However, this result was attained by not taking into consideration certain external factors that Loïck Peyron may have encountered when he broke the record. Therefore, if the condition were just right, and a constant speed of 40 knots was kept consistent throughout, the results from this examination would be valid.

Word Count: 300

Table of Contents

A. Acknowledgment

Before we can attempt to hypothetically break a record, some distinctions have to be addressed. Under no circumstances, is this examination defaming the current record holder, Loïck Peyron, by providing an alternative route, or method. External factors such as climate, the Earth not being perfectly spherical(radius of 6,371 Km will be used) and boundaries will not be taken into consideration, due to the uncertainties they may implicate in the examination; average speed of sailboat under regular circumstances will be used. This examination, is only a hypothetical attempt; the routes presented during this examination have not been executed in the real world or will ever be by the author of this examination.

B. Introduction

Euclid is the father of geometry and, though he might not have been the one who discovered it, he is accredited for it because he develop the first comprehensive deductive system. Non-Euclidean geometries are based on Euclid's postulates, but respectively use their own version of the parallel, fifth, postulate. Non-Euclidean Geometry is accredited to four people: C.F. Gauss (1777-1855) for discovering the possibility of Non-Euclidean geometry, N. Lobachevsky (1792-1856) for the discovery of hyperbolic geometry, J. Bolyai (1802-1860) for adding onto Gauss' and Lobachevsky's work and B. Riemann (1826-1866) for his development in spherical, or elliptical, geometry. For this examination the sub-branch of Non-Euclidean geometry, spherical trigonometry, will be used to, **Propose a Hypothetical Method of Attempting to Break the Current Sailing Record Around the World using Spherical Trigonometry.**

In the world, especially on the surface of the Earth, lines are not always straight. For example if an airplane from California was heading to China, the route it would take would not be a straight line; if so, the airplane would have to cut through the Earth. Spherical Trigonometry, is a sub-branch of Bernhard Riemann's spherical geometry. Just like in Euclidean geometry there is trigonometry, there is also for when the plane is spherical. Not just does this implicate that there is a difference in formulas, but also how problems will be approached.

Before the attempt of hypothetically breaking a world record, it is important to analyze and study the record holder's, Loïck Peyron, route, equipment, speed etc. Loïck Peyron, in his world-record-breaking run, used a Zoulou Extreme 40, which has an average speed of 40 knots(~46 mph, 74 km/h). The route taken by Loïck Peyron, began and ended in an imaginary line between the Créac'h lighthouse on Ouessant (Ushant) Island, France, and the Lizard Lighthouse, UK located on the coordinates 48°27'34.23"N, 5°7'45.4"W. It took him 45 days 13 hours 42 minutes and 53 seconds to break the record, hence this is the time to beat.

C. Spherical Trigonometry

It takes basic knowledge in geometry to know that the shortest distance between two points is a straight line; but on the surface of a sphere, there are no such things as straight lines. The shortest distance between two points on a sphere is the arc of a *great circle* passing through those points. "A great circle is defined to be the intersection with a sphere on a plane containing the center of the sphere, [such that of figure-1 and 2]. If the plane does not contain the center of the sphere, its intersection with the sphere is known as a small circle, [such that of figure-3 and 4]"(Dhillon).

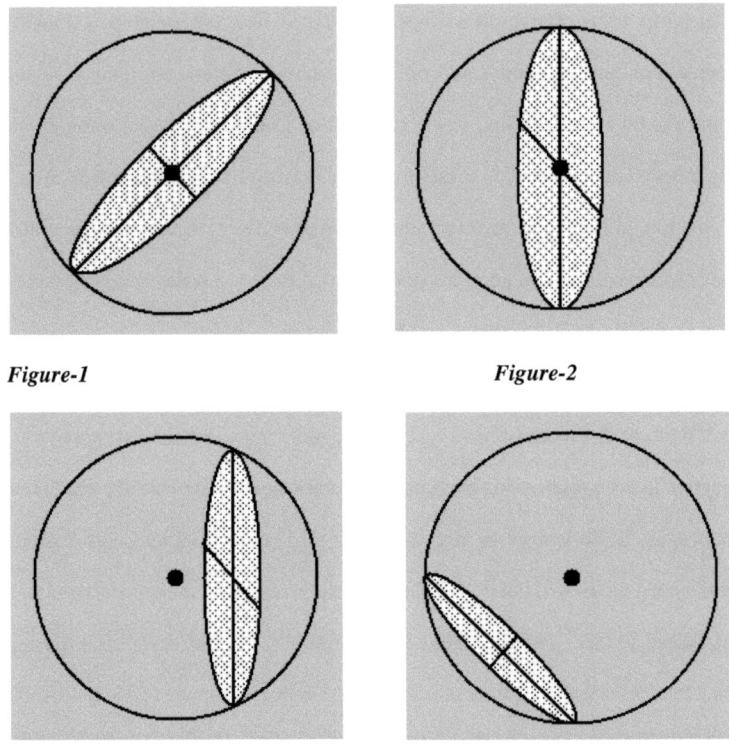

Figure-1   *Figure-2*

Figure-3   *Figure-4*

The main concept that this investigation will deal with, is with the triangle; more specifically, spherical triangles. In Euclidean geometry when we connect three points on a plane using the shortest possible route, it will create a triangle. By analogy, in Non-Euclidean geometry when we want to connect three points on the surface of a sphere, "we would draw arcs of great circles and hence create a spherical trian-gle"(Dhillon). Because spherical triangles do not necessarily have to look like planar triangles, a triangle on the surface of a sphere is only a spherical triangle if *all* the fol-lowing properties are true: the three sides are all arcs of great circles, any two sides are together longer than the third side, the sum of the three angles is greater than 180°

(π radians), and if each individual spherical angle is less than 180°.

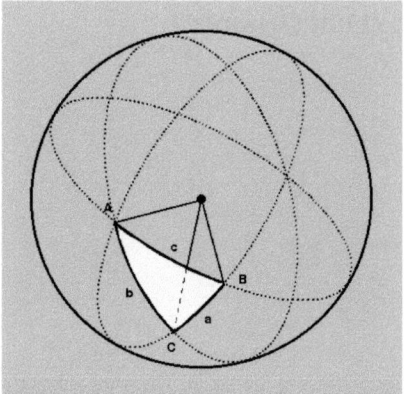

Figure-5

As this examination will deal with the surface of the Earth, it is important to know how spherical trigonometry can be expressed. Two coordinates, latitude and longitude, are used to pin-point any location on the surface of the Earth. The longitude of a point, is measured angularly from east to west of the Greenwich Meridian(great circle that passes through both poles of the Earth) with respect to the equator. If the longitude is east of the Greenwich Meridian, then it is a positive angle. If the longitude is west of the Greenwich meridian, then it is a negative angle. The latitude of a point, is measured angularly from north to south of the equator with respect to the Greenwich Meridian. If the latitude is north of the equator, then the angle is positive. If the latitude is is south of the equator, then the angle will be negative.

Laws pertaining to Euclidean geometry, more specifically trigonometry, still exist in Non-Euclidean. Laws like the law of sine and cosine still, accurately, can be used to solve problems. There are some distinction the law of cosine, as it does not look entirely like the Non-Euclidean version, but the law of sine is extremely similar. Where print letters are the sides of the triangle and roman being the interior angles.

The law of Cosine is represented as such:

$$\cos(a) = \cos(b)\cos(c) + \cos(\alpha)\sin(b)\sin(c)$$

The law of Sine is represented as such:

$$\frac{\sin(a)}{\sin(\alpha)} = \frac{\sin(b)}{\sin(\beta)} = \frac{\sin(c)}{\sin(\gamma)}$$

D. Record Breaking Attempt

To begin, we must first decide on the route that will be taken. The route taken by Loïck Peyron, began and ended in an imaginary line between the Créac'h lighthouse on Ouessant (Ushant) Island, France, and the Lizard Lighthouse, UK located on the coordinates 48°27′34.23″N, 5°7′45.4″W. Latitude and Longitude's standard notation is sexagesimal (shown above), but for the purpose of this examination, they will be changed into decimal degrees, and then into radians.

The first step in doing so, is to change the seconds portion of the latitude or longitude into a decimal of minutes. The second step, is to set the minutes over sixty and it equal to "x". Next, add the "x" value to the degrees. Lastly, multiply the degrees by π ,and divide by 180. For example, the coordinates of the Créac'h lighthouse on Ouessant (Ushant) Island, France, and the Lizard Lighthouse, UK located on the coordinates 48°27′34.23″N, 5°7′45.4″W will be changed into decimal degrees, then into radians. Varying, wether the latitude and longitude are going north, south, east or west put the correct + or - sign; as this will be crucial in the subtraction.

$$34.23 / 60 = 0.5705$$
$$0.5705 + 27 = 27.5705$$
$$27.5705 / 60 = 0.4595$$
$$x = 0.4595$$
$$48 + x = 48.4595°$$
$$(48.4595)(\frac{\pi}{180}) = 0.8473$$

This process will be repeated with the rest of the latitudes and longitudes of the points remaining. Table-1 displays the given points of reference to guide the boat , with their respective latitude and longitude in both sexagesimal and decimal degree form (turned into radians).

The selection process for the points chosen was not a specific one. I merely looked at a globe with the destination and pin pointed points what seemed to me the most reasonable, yet shortest to arrive at the destination. Figure - 7 represents the points on Table - 1 plotted.

	Lat/Sexagesimal	Lat/Decimal	Lon/Sexagesimal	Lon/Decimal
Start/End	48°27'34.23"N	0.8457	5°7'45.4"W	0.0895
1	24° 45' 11.97"N	0.432	42° 10' 25.23"W	0.7361
2	12° 1' 21.30"N	0.2098	61° 40' 43.71"W	1.0765
3	17° 59' 32.52"N	0.314	76° 47' 30.91"W	1.3403
4	9° 21' 30.88"N	0.1633	79° 54' 0.05"W	1.3945
5	9° 3' 26.65"N	0.1581	79° 39' 28.54"W	1.3903
6	8° 15' 42.80"N	0.1442	79° 6' 31.25"W	1.3807
7	4° 59' 38.53"N	0.0872	77° 12' 35.42"W	1.3476
8	20° 28' 0.81"N	0.3572	156° 29' 38.22"W	2.7837
9	70° 40' 31.64"N	1.2335	23° 51' 13.34"E	-0.4163
10	50° 16' 10.19"N	0.8774	19° 56' 17.52"W	0.348

Table-1

* plotted on figure-6

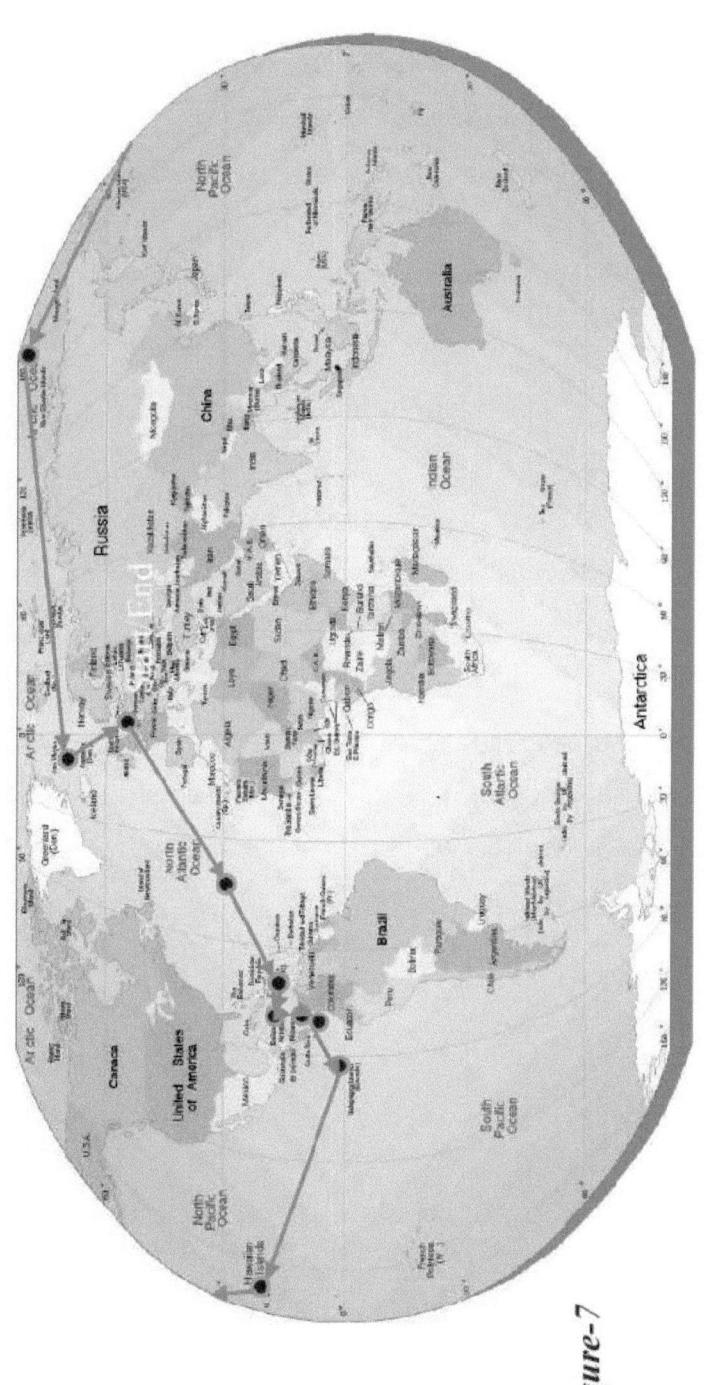

Figure-7

*Not to scale

After all the coordinates have been found, the Haversine formula will be applied to find the distance between 2 distinct latitudes and longitudes

This will be the formula that will be used with most frequency throughout this examination:

Haversine Formula: Used to find the distance between 2 distinct Latitudes and Longitudes.

$$hav(\frac{d}{r}) = \sin^2(\frac{dlat}{2}) + \cos(lat_1)\cos(lat_2)\sin(\frac{dlon}{2})$$

$$hav(\frac{d}{r}) = \sin^2(\frac{d}{2r})$$

Where dlat = difference of latitudes, dlon = difference of longitude, lat_1, and lat_2 are the latitudes of their respective coordinate, d = distance of great circle arc, and

r = radius of earth. All angles will be measured in radians.

This examination will begin with the usage of the haversine formula to find the length between start point and point #1.

$$hav(\frac{d}{r}) = \sin^2(\frac{dlat}{2}) + \cos(lat_1)\cos(lat_2)\sin^2(\frac{dlon}{2})$$

$$dlat = lat_2 - lat_1$$

$$lat_1 = 0.8457, lat_2 = 0.432$$

$$dlat = 0.432 - 0.8457$$

$$dlat = -0.4137$$

$$dlon = lon_2 - lon_1$$

$$lon_1 = 0.0895, lon_2 = 0.7361$$

$$dlon = 0.7361 - 0.0895$$

$$dlon = 0.6466$$

$$r = 6371 Km$$

$$hav(\frac{d}{6371}) = \sin^2(\frac{-0.4137}{2}) + \cos(0.8457)\cos(0.432)\sin^2(\frac{0.6466}{2})$$

$$hav(\frac{d}{6371 Km}) = 0.10297$$

$$\sin^2(\frac{d}{(2)6371 Km}) = 0.10297$$

$$\sin(\frac{d}{12742 Km}) = \sqrt{0.10297}$$

$$\frac{d}{12742 Km} = \arcsin(0.32089)$$

$$d = 4162.0631 Km \rightarrow a$$

Once the first distance, from starting point to point #1, has been attained, the next objective is to find the distance between point #1 and #2.

$$hav(\frac{d}{r}) = \sin^2(\frac{dlat}{2}) + \cos(lat_1)\cos(lat_2)\sin^2(\frac{dlon}{2})$$

$dlat = lat_2 - lat_1$

$lat_1 = 0.432, lat_2 = 0.2098$

$dlat = 0.2098 - 0.432$

$dlat = -0.2222$

$dlon = lon_2 - lon_1$

$lon_1 = 0.7361, lon_2 = 1.0765$

$dlon = 1.0765 - 0.7361$

$dlon = 0.3404$

$r = 6371 Km$

$$hav(\frac{d}{6371Km}) = \sin^2(\frac{-0.2222}{2}) + \cos(0.432)\cos(0.2098)\sin^2(\frac{0.3404}{2})$$

$$hav(\frac{d}{6371Km}) = 0.03777$$

$$\sin^2(\frac{d}{(2)6371Km}) = 0.03777$$

$$\sin(\frac{d}{12742Km}) = \sqrt{0.03777}$$

$$\frac{d}{12742Km} = \arcsin(0.19436)$$

$d = 2492.3691 Km \rightarrow b$

The next objective, is to find the length between point #2 and point #3.

$$hav(\frac{d}{r}) = \sin^2(\frac{dlat}{2}) + \cos(lat_1)\cos(lat_2)\sin^2(\frac{dlon}{2})$$

$$dlat = lat_2 - lat_1$$

$$lat_1 = 0.2098, lat_2 = 0.314$$

$$dlat = 0.314 - 0.2098$$

$$dlat = 0.1042$$

$$dlon = lon_2 - lon_1$$

$$lon_1 = 1.0765, lon_2 = 1.3403$$

$$dlon = 1.3403 - 1.0765$$

$$dlon = 0.2638$$

$$r = 6371Km$$

$$hav(\frac{d}{6371}) = \sin^2(\frac{0.1042}{2}) + \cos(0.2098)\cos(0.314)\sin^2(\frac{0.2638}{2})$$

$$hav(\frac{d}{6371Km}) = 0.0188$$

$$\sin^2(\frac{d}{(2)6371Km}) = 0.0188$$

$$\sin(\frac{d}{12742Km}) = \sqrt{0.0188}$$

$$\frac{d}{12742Km} = \arcsin(0.13712)$$

$$d = 1752.7049Km \rightarrow c$$

The next objective, is to find the length between point #3 and point #4.

$$hav(\frac{d}{r}) = \sin^2(\frac{dlat}{2}) + \cos(lat_1)\cos(lat_2)\sin^2(\frac{dlon}{2})$$

$dlat = lat_2 - lat_1$

$lat_1 = 0.314, lat_2 = 0.1633$

$dlat = 0.1633 - 0.314$

$dlat = -0.1507$

$dlon = lon_2 - lon_1$

$lon_1 = 1.3404, lon_2 = 1.3945$

$dlon = 1.3945 - 1.3404$

$dlon = 0.0541$

$r = 6371Km$

$$hav(\frac{d}{6371}) = \sin^2(\frac{-0.1507}{2}) + \cos(0.1633)\cos(0.314)\sin^2(\frac{0.05}{2}$$

$$hav(\frac{d}{6371Km}) = 0.00635$$

$$\sin^2(\frac{d}{(2)6371Km}) = 0.00635$$

$$\sin(\frac{d}{12742Km}) = \sqrt{0.00635}$$

$$\frac{d}{12742Km} = \arcsin(0.07969)$$

$d = 1016.4878Km \rightarrow d$

The next objective, is to find the length between point #4 and point #5.

$$hav(\frac{d}{r}) = \sin^2(\frac{dlat}{2}) + \cos(lat_1)\cos(lat_2)\sin^2(\frac{dlon}{2})$$

$$dlat = lat_2 - lat_1$$

$$lat_1 = 0.1633, lat_2 = 0.1581$$

$$dlat = 0.1581 - 0.1633$$

$$dlat = -0.0052$$

$$dlon = lon_2 - lon_1$$

$$lon_1 = 1.3945, lon_2 = 1.3903$$

$$dlon = 1.3903 - 1.3945$$

$$dlon = -0.0042$$

$$r = 6371Km$$

$$hav(\frac{d}{6371}) = \sin^2(\frac{-0.0052}{2}) + \cos(0.1633)\cos(0.1581)\sin^2(\frac{-0.0042}{2})$$

$$hav(\frac{d}{6371Km}) = 0.000011$$

$$\sin^2(\frac{d}{(2)6371Km}) = 0.000011$$

$$\sin(\frac{d}{12742Km}) = \sqrt{0.000011}$$

$$\frac{d}{12742Km} = \arcsin(0.00332)$$

$$d = 42.3035Km \rightarrow e$$

The next objective, is to find the length between point #5 and point #6.

$$hav(\frac{d}{r}) = \sin^2(\frac{dlat}{2}) + \cos(lat_1)\cos(lat_2)\sin^2(\frac{dlon}{2})$$

$$dlat = lat_2 - lat_1$$

$$lat_1 = 0.1581, lat_2 = 0.1442$$

$$dlat = 0.1442 - 0.1581$$

$$dlat = -0.0139$$

$$dlon = lon_2 - lon_1$$

$$lon_1 = 1.3903, lon_2 = 1.3807$$

$$dlon = 1.3807 - 1.3903$$

$$dlon = -0.0096$$

$$r = 6371Km$$

$$hav(\frac{d}{6371}) = \sin^2(\frac{-0.0139}{2}) + \cos(0.1581)\cos(0.1442)\sin^2(\frac{-0.0}{2}$$

$$hav(\frac{d}{6371Km}) = 0.000071$$

$$\sin^2(\frac{d}{(2)6371Km}) = 0.000071$$

$$\sin(\frac{d}{12742Km}) = \sqrt{0.000071}$$

$$\frac{d}{12742Km} = \arcsin(0.00843)$$

$$d = 107.4163Km \rightarrow f$$

The next objective, is to find the length between point #6 and point #7.

$$hav(\frac{d}{r}) = \sin^2(\frac{dlat}{2}) + \cos(lat_1)\cos(lat_2)\sin^2(\frac{dlon}{2})$$

$dlat = lat_2 - lat_1$

$lat_1 = 0.1442, lat_2 = 0.0872$

$dlat = 0.0872 - 0.1442$

$dlat = -0.057$

$dlon = lon_2 - lon_1$

$lon_1 = 1.3807, lon_2 = 1.3476$

$dlon = 1.3476 - 1.3807$

$dlon = -0.0331$

$r = 6371Km$

$$hav(\frac{d}{6371}) = \sin^2(\frac{-0.057}{2}) + \cos(0.1442)\cos(0.0872)\sin^2(\frac{-0.0331}{2})$$

$$hav(\frac{d}{6371Km}) = 0.00108$$

$$\sin^2(\frac{d}{(2)6371Km}) = 0.00108$$

$$\sin(\frac{d}{12742Km}) = \sqrt{0.00108}$$

$$\frac{d}{12742\,Km} = \arcsin(0.03286)$$

$d = 418.7775Km \rightarrow g$

The next objective, is to find the length of point #7 and point #8.

$$hav(\frac{d}{r}) = \sin^2(\frac{dlat}{2}) + \cos(lat_1)\cos(lat_2)\sin^2(\frac{dlon}{2})$$

$$dlat = lat_2 - lat_1$$

$$lat_1 = 0.0872, lat_2 = 0.3572$$

$$dlat = 0.3572 - 0.0872$$

$$dlat = 0.27$$

$$dlon = lon_2 - lon_1$$

$$lon_1 = 1.3476, lon_2 = 2.7837$$

$$dlon = 2.7837 - 1.3476$$

$$dlon = 1.4361$$

$$r = 6371 Km$$

$$hav(\frac{d}{6371}) = \sin^2(\frac{-0.0139}{2}) + \cos(0.1581)\cos(0.1442)\sin^2(\frac{-0.0}{2}$$

$$hav(\frac{d}{6371 Km}) = 0.4221$$

$$\sin^2(\frac{d}{(2)6371 Km}) = 0.4221$$

$$\sin(\frac{d}{12742 Km}) = \sqrt{0.4221}$$

$$\frac{d}{12742 Km} = \arcsin(0.64969)$$

$$d = 9010.8439 Km \rightarrow h$$

The next objective, is to find the length of point #8 and point #9.

$$hav(\frac{d}{r}) = \sin^2(\frac{dlat}{2}) + \cos(lat_1)\cos(lat_2)\sin^2(\frac{dlon}{2})$$

$$dlat = lat_2 - lat_1$$

$$lat_1 = 0.3572, lat_2 = 1.2335$$

$$dlat = 1.2335 - 0.3572$$

$$dlat = 0.8763$$

$$dlon = lon_2 - lon_1$$

$$lon_1 = 2.7837, lon_2 = -0.4163$$

$$dlon = -0.4163 - 2.7837$$

$$dlon = -3.2$$

$$r = 6371Km$$

$$hav(\frac{d}{6371}) = \sin^2(\frac{0.8763}{2}) + \cos(1.2335)\cos(0.3572)\sin^2(\frac{-3.2}{2})$$

$$hav(\frac{d}{6371Km}) = 0.48978$$

$$\sin^2(\frac{d}{(2)6371Km}) = 0.48978$$

$$\sin(\frac{d}{12742Km}) = \sqrt{0.48978}$$

$$\frac{d}{12742Km} = \arcsin(0.69984)$$

$$d = 9877.2604Km \rightarrow i$$

The next objective, is to find the length between point #9 and point #10.

$$hav(\frac{d}{r}) = \sin^2(\frac{dlat}{2}) + \cos(lat_1)\cos(lat_2)\sin^2(\frac{dlon}{2})$$

$$dlat = lat_2 - lat_1$$

$$lat_1 = 1.2335, lat_2 = 0.8774$$

$$dlat = 0.8774 - 1.2335$$

$$dlat = -0.3561$$

$$dlon = lon_2 - lon_1$$

$$lon_1 = -0.4163, lon_2 = 0.348$$

$$dlon = 0.348 + 0.4163$$

$$dlon = 1.6398$$

$$r = 6371 Km$$

$$hav(\frac{d}{6371}) = \sin^2(\frac{-0.3561}{2}) + \cos(1.2335)\cos(0.8774)\sin^2(\frac{0.7\epsilon}{2}$$

$$hav(\frac{d}{6371 Km}) = 0.20941$$

$$\sin^2(\frac{d}{(2)6371 Km}) = 0.20941$$

$$\sin(\frac{d}{12742 Km}) = \sqrt{0.20941}$$

$$\frac{d}{12742 Km} = \arcsin(0.45761)$$

$$d = 6056.3412 Km \rightarrow j$$

The next objective, is to find the length between point #10 and end point.

$$hav(\frac{d}{r}) = \sin^2(\frac{dlat}{2}) + \cos(lat_1)\cos(lat_2)\sin^2(\frac{dlon}{2})$$

$dlat = lat_2 - lat_1$

$lat_1 = 0.8774, lat_2 = 0.8457$

$dlat = 0.8457 - 0.8774$

$dlat = -0.0317$

$dlon = lon_2 - lon_1$

$lon_1 = 0.348, lon_2 = 0.0895$

$dlon = 0.0895 - 0.348$

$dlon = -0.2585$

$r = 6371 Km$

$$hav(\frac{d}{6371}) = \sin^2(\frac{-0.0317}{2}) + \cos(0.8774)\cos(0.8457)\sin^2(\frac{-0.2585}{2})$$

$$hav(\frac{d}{6371 Km}) = 0.00729$$

$$\sin^2(\frac{d}{(2)6371 Km}) = 0.00729$$

$$\sin(\frac{d}{12742 Km}) = \sqrt{0.00729}$$

$$\frac{d}{12742 Km} = \arcsin(0.08538)$$

$d = 1089.238 Km \rightarrow k$

Once all the lengths have been attained, it is time to find out how many days, hours, minutes, and seconds its going to take to elapse that distance. We know that the average speed of the boat used by Loïck Peyron is of 40 knots. For this investigation, the unit of measurement used all along was that of kilometers,therefore when finding the time, we will measure the 40knots as 74Km/h.

$$a+b+c+d+e+f+g+h+i+j+k$$
$$a+b \Rightarrow 4162.0631\,Km + 2492.3691\,Km = 6654.4322\,Km$$
$$c+d \Rightarrow 1752.7049\,Km + 1016.4878\,Km = 2769.1927\,Km$$
$$e+f \Rightarrow 42.3035\,Km + 107.4163\,Km = 149.7198\,Km$$
$$g+h \Rightarrow 418.7775\,Km + 9010.8439\,Km = 9429.6214\,Km$$
$$i+j+k \Rightarrow 9877.2604\,Km + 6056.3412\,Km + 1089.238\,Km = 17022.839$$
$$a+b+c+d+e+f+g+h+i+j+k = 36025.8057\,Km$$

$$\frac{36025.8057\,Km}{\dfrac{74km}{hr}} = 36025.8057\,Km\left(\frac{1hr}{74\,Km}\right)$$

$$36025.8057\,Km\left(\frac{1hr}{74\,Km}\right) = 486.8352hr$$

$$\frac{486.8352hr}{24hr} = 20.2848\,Days$$

$$\frac{2848}{10000} = \frac{Hours}{60}$$

$$Hours = 17.088$$

$$\frac{088}{1000} = \frac{Minutes}{60}$$

$$Minutes = 5.28$$

$$\frac{28}{100} = \frac{Seconds}{60}$$

$$Seconds = 16.8 \approx 17$$

We end up with a final time 20 days, 17 hours, 5 minutes, and 17 seconds.

E. Conclusion

Through this investigation the statement, **Propose a Hypothetical Method of Attempting to Break the Current Sailing Record Around the World using Spherical Trigonometry**, was challenged. The examination included a potential method of sailing around the world in an attempt to break the current record held by Loïck Peyron, in 45 days 13 hours 42 minutes and 53 seconds. The path that Loïck Peyron took on his record-breaking run was inaccessible, so there was no comparison to be made with the path taken in this examination. Table-1 and Figure-6 show the path in which this examination was based. There is no specific relevance as to why those points were chosen, other than seeming to be the shortest path to get around the world and ending in the Créac'h lighthouse on Ouessant (Ushant) Island, France, and the Lizard Lighthouse, UK located on the coordinates 48°27'34.23"N, 5°7'45.4"W.

The result of this examination ended with respect to the points on Table-1 was that the record was broken using this particular method.The time elapsed to complete the whole route was exactly 20 days, 17 hours, 5 minutes, and 17 seconds. Through the use of the haversine formula, the distance between the points represented in Table-1 were found to ten-thousandths.Followed by their sum and converting the Km time by taking into consideration Loïck Peyron's boat speed. It must also be taken into consideration, that external factors such as climate, not always maintain the speed of 40 knots and the Earth not being perfectly spherical, were not taken into consideration. Henceforth, altering the results from this examination. It could be, that if these other factors were taken into consideration, the path taken in this examination would have not stood up Loïck Peyron's record.

25

Another aspect that can be explored with this same concept, is celestial navigation. It is very similar to that of this examination, as it roots to Non-Euclidean geometry much like how spherical trigonometry does. Even this very proposal:Propose a Hypothetical Method of Attempting to Break the Current Sailing Record Around the World using Spherical Trigonometry, can be challenged by approaching it with an aspect of celestial navigation. The only difference would be, that instead of finding latitudes and longitudes, you would have to measure distance by the location of stars at a given time and position, making it more complex and thought-provoking.

F. Works Cited

Dhillon, Vik. *Spherical Trigonometry*. Digital image. *Sheffield University*. Web. <http://vikdhillon.staff.shef.ac.uk/teaching/phy105/celsphere/phy105_trigonomet ry.html>.

Dhillon, Vik. *Spherical Trigonometry*. Sheffield: Sheffield University.

"Euclid." Encyclopedia Britannica. 8th ed. 2009. Print.

Gibilisco, Stan. "Global Trigonometry." Trigonometry Demystified. New York: McGraw-Hill, 2003. 224-46. Print.

Greenberg, Marvin J. "The Discovery of Non-Euclidean Geometry." Euclidean and Non-Euclidean Geometries: Development and History. San Francisco: W.H. Freeman, 1980. 140-60. Print.

Hedges, Andrew. "Finding Distances Based on Latitude and Longitude." Web. <http://andrew.hedges.name/experiments/haversine/>.

Weisstein, Eric W. "Haversine." *Wolfram Wathworld*. Wolfram. Web. <http://mathworld.wolfram.com/Haversine.html>.

World Map. Digital image. *The True News*. Web. <http://thetruthnews.info/world_map.gif>.

YOUR KNOWLEDGE HAS VALUE

- We will publish your bachelor's and
  master's thesis, essays and papers

- Your own eBook and book -
  sold worldwide in all relevant shops

- Earn money with each sale

Upload your text at www.GRIN.com
and publish for free